D0623485

ISABELLA BREGA

Canyon Country

THE STONE GARDENS OF THE WEST

CANYON COUNTRY

CONTENTS

Text
Isabella Brega

Graphic design
Patrizia Balocco

Editorial coordination
Valeria Manferto De Fabianis
Alberto Bertolazzi

Translation
Ann Hylands Ghiringhelli

© 1995 White Star S.r.l.
Via Candido Sassone, 22/24
13100 Vercelli, Italy.

All rights reserved. No part of this publication may be reproduced, stored in a retrieval system or transmitted in any form by any means electronic, mechanical, photocopying or otherwise, without first obtaining written permission of the copyright owner.

This edition published in 1995 by Smithmark Publishers, a division of U.S. Media Holdings, Inc., 16 East 32nd Street, New York, NY 10016.

SMITHMARK books are available for bulk purchase, for sales promotion and premium use. For details write or call the manager of special sales, SMITHMARK Publishers, 16 East 32nd Street, New York, NY 10016; (212) 532-6600.

Produced by: White Star S.r.l.
Via Candido Sassone, 22/24
13100 Vercelli, Italy.

ISBN: 0-8317-1003-9

Printed in July 1995 by Grafedit, Bergamo, Italy.

1 To the south of the Natural Bridges National Monument is the Valley of the Gods, an example of how — over a period of hundreds of thousands of years — the eroding force of the elements has shaped the landscape of America's West.

2-3 Especially in the states of the West, Utah and Arizona included, the national parks of America turn nature into spectacle, offering breathtaking sights — in Monument Valley, for instance — that have provided the backdrop for countless movies and commercials.

4-5 According to an ancient Indian legend, Bryce Canyon is formed of animals which had the power to turn themselves into men. But Shin-Owav, a demigod of the Paiute, punished them for their mischief by transforming them into rocks.

6-7 The Grand Canyon National Park, in Arizona, is like an open textbook on the geological history of our planet. Each layer of rock, each different colour, corresponds to a different epoch and eco-system. Within the confines of the park are six of the Earth's seven climatic zones.

8 Bryce Canyon National Park consists of a gigantic amphitheatre with walls of rock, hollowed out by wind and rain from the side of the Paunsaugunt Plateau; in the language of the Paiute Indians Paunsaugunt means "home of the beaver".

9 Created by millions of years of abrasion by wind and water on layers of sandstone, the amazing rock sculptures of Arches National Park form a timeless wonderland which draws hundreds of thousands of visitors every year.

10-11 "One hell of a place to lose a cow" is how Bryce Canyon was described by the man it was eventually called after. A Mormon settler, Ebenezer Bryce came here with his wife in 1875 to raise cattle but five years in this hostile place was the limit of their resistance.

Introduction

If you want to test the truth of the Latin saying "gutta cavat lapidem" (a drop hollows out stone), go to the American West — especially the states of Utah and Arizona — and contemplate the awesome expanses of a still sparsely populated land where nature has in many respects remained intact. Circling the huge Navajo reservation — America's largest, occupying, with the Hopi reservation, about a quarter of Arizona's land area as well as part of Colorado, Utah and New Mexico — you will find the extraordinary national parks of Grand Canyon, Bryce Canyon, Arches, Petrified Forest, Canyonlands and Monument Valley, the national monuments of Natural Bridges and Canyon De Chelly, the awe-inspiring gorges of Mesa Verde and Paria. These marvels of nature were created by complex geological phenomena, movements of the Earth's crust and, above all, millions of years of erosion by water. During the eons of prehistory the waters of the Colorado river, the ice of bitterly cold winters and driving rain wore down, hollowed out, crushed, pulverized, smoothed and shaped these arid sandstone tablelands, formed of rock left by primordial sedimentation. With the passing days, years and centuries, the combined forces of water and wind created an amazing assortment of eroded rock formations and landscapes that have no equal on this planet. Here time has produced another natural wonder in the form of petrified forests: silicified remains of trees that grew on wooded, green plains more than 200 million years ago and have returned to the surface of what is now a barren desert. On these plateaus of the West, the America we associate with youth, modernity and progress, a country that has come to symbolize the future, reveals its "old continent" past and monumental features of a size and splendour unknown in Europe. Rock formations and the effects of erosion and weathering are the key to Canyon Country: the mighty gorge of the Grand Canyon,

12-13 With its vast expanse of blue, Lake Powell — the huge artificial lake that occupies the heart of Glen Canyon, in Arizona — is like an oasis in the midst of a desert. Carp and perch are among the many fish that now populate its waters.

the layered pillars of Bryce Canyon, the splendid rock sculptures of Arches National Park and Natural Bridges, the imposing mesas of Monument Valley and Canyonlands. And age-old pueblos and rock paintings prove that this land was populated by men — migrated from Asia when there was still a natural bridge between the two continents — long before its "discovery" by the Spaniards. In this primordial, barren, pitiless land nature still exists in the raw. But her crueler manifestations are offset by the all-embracing calm of the first rays of dawn, the stirring sight of flaming sunsets and the serenity of horizons that fade into infinity. The process of erosion by water and wind continues, slowly but surely — even if imperceptibly — changing the landscape of the West. No photo, movie or documentary has ever succeeded in fully conveying the breathtaking splendour of this territory which has attracted people throughout the centuries: from the mysterious Anasazi to pioneers and adventurers and, in our own times, film directors like John Ford and Sergio Leone, who gave shape to the legends of the West. Sites of movie fiction like Monument Valley or geological facts like the Grand Canyon have been 'taken over' by white men. They nonetheless belong by birthright to the native Indians, many of whom live in the reservations that border on or, in some cases, form part of the parks. Navajo, Hopi, Havasupai and Hualapai: legendary names of peoples who are now captives of their stereotyped image, prisoners not of boundaries but of their glorious past and tragic defeat. Each year millions of visitors come here to pass in review rocks that tell the geological history of our planet, and stare spellbound at their gigantic dimensions, changing colours and extraordinary wealth of bizarre shapes. This is a land to be experienced first-hand by succumbing to the irresistible call of adventure or by simply contemplating its wonders, peering giddily down into its depths or looking intently towards the great beyond to see where — if ever — the America epitomized by Canyon Country comes to an end.

14-15 Tiny multi-coloured splinters of petrified wood, enormous tree-trunks transformed into quartz crystals, amid the colourful hills and mesas of the badlands: the Petrified Forest National Park is part of a barren, desert-like area of northeastern Arizona.

16-17 Upriver from the Grand Canyon, the dam built in Glen Canyon serves to control the waters of the Colorado and form the reservoir that now goes by the name of Lake Powell. Besides taming the flow of the river, however, the scheme has lowered water temperature, bringing changes in the species of fish that make the river their habitat.

18-19 Monument Valley's towering mesas, deep gorges, natural bridges and arches, sandstone pinnacles and monoliths are all part of the vast Navajo Indian reservation, the largest in the USA.

GRAND CANYON NATIONAL PARK

We have all heard it described as one of the world's greatest natural wonders. We have seen it in hundreds of films, documentaries and photos and yet, even when it is stretched out before us, we are not quite ready to cope with this experience. A first glance at the Grand Canyon from the edge of the South Rim leaves visitors speechless. However many descriptions we have read and pictures we have seen, this dizzying, stratified chasm revealing rocks almost as old as the planet itself is beyond all imagination: a streaked red wound which plunges downwards, through layer after layer, cutting almost 1600 metres into the heart of the Earth — unexpected, exhilarating and disquieting. With a length of 445 kilometres, a maximum width of 30 kilometres and an area of almost 5000 square kilometres, the Grand Canyon National Park in northern Arizona is absolutely unique. As you peer into the depths of the gorge, its walls offer an overview of almost two billion years of geological events (though geologists have still to agree on "exact" dates). Each rock stratum, each colour, corresponds to a different epoch and ecosystem. In all there are twelve rock formations — volcanic, limestone, sandstone, granite — and the oldest deposits date back to the Paleozoic era: at the very bottom of the canyon is shiny black Vishnu Schist, one of the Earth's primordial rocks from the Pre-Cambrian period; the next layers are formed of greenish Muav limestone and blue-grey Redwall, both stemming from marine deposits and with abundant traces of ferns and other lifeforms. Close to the surface there is Coconite sandstone, of sand dune origin, and, on the very top, Kaibab grey limestone, containing fossils from the period when this entire region was covered by sea tens of millions of years ago. And this rock formation, the youngest of the Grand Canyon, was here before dinosaurs trod the Earth... If present-day visitors —

20-21 Along the South Rim of the Grand Canyon are twelve lookout points from which to peer into the abyss below or gaze in wonder at the general panorama. The most spectacular is perhaps Grandview Point while Mather Point offers the most famous vista of the Canyon; other breathtaking views are to be had from Hopi Point, Mohavi Point (the best two sites from which to scan the winding course of the Colorado), Navajo Point and the stunning Bright Angel Point.

accustomed to special effects "à la Spielberg" — are left awestruck by the Grand Canyon, what impact must it have made on the first white men who chanced this way... When searching for the legendary Golden Cities of Cibola in 1540, a Spanish officer, Garcia Lopez de Cardonas, was overwhelmed by the sight of this wonder of nature, with "mountains higher than the cathedral of Seville". Years later another Spaniard, the Franciscan missionary Francisco Tomas Garcés, came across the Grand Canyon and its wild river which, given its reddish colour, this man of the cloth "baptized" with the name of Rio Colorado. This great, winding river is surely the protagonist of the Grand Canyon and its park. Its flow has now been tamed by two dams but tens of millions of years ago its unruly waters worked their way through the Kaibab limestone of the plateau — once the bed of an ocean and now, due to movements of the Earth's crust, 1500 metres above sea level — and then through the rock strata below, to eventually create the Canyon. With almost seventy rapids along its course, the Colorado River offers a journey of adventure and discovery that many men have been tempted to undertake. The now legendary figure most often associated with the river is Major John Wesley Powell, a Civil War veteran who had lost an arm but not the determination to challenge the unknown; he made his first voyage down the Colorado River in 1869, and a second in 1872. Attempts had been made even earlier in 1826 by a trapper, James Ohio Pattie, and in 1856 by Joseph Christmas Ives, whose steam ship Explorer was halted by the rocks of Black Canyon. For centuries the Indians had been lords of the canyon. There are traces of a mysterious people who, around 2000 BC, left crudely fashioned and engraved statuettes in a cave here. Later, about 500 AD, the region was populated by Anasazi, who survived on the produce of tiny plots of land, pinon pine seeds and cactus fruit. And there are still Indians living here today: a small community of

22 Within the territory of the Grand Canyon, made a national monument by President Roosevelt in 1908 and a national park in 1919, is a uranium mine that, until the sixties, had more abundant ore deposits than any other in the USA.

23 Descending deeper towards the floor of the canyon, where less rain falls and the temperature is higher, the vegetation changes. Among the flora found at Tonto Platform, about 1000 metres above sea level, is the Utah century plant, which reaches maturity and flowers only after many years (below). Like all plants of the agave genus, it flowers just once and then dies.

24-25 The North Rim is 300 metres or so higher than the South Rim and the temperature on the two sides of the canyon consequently differs. In summer, for instance, the temperature on the southern side is around 30°C as compared with 24°C on the northern side. On the floor of the canyon, it can instead easily go above 40°C. There is a difference in annual rainfall too: about 700 mm on the North Rim, only 400 mm on the South Rim and less than 200 mm in the canyon itself. Likewise, in winter snow can reach a depth of 3 metres on the North Rim (which is closed to visitors from the end of October to mid-May) while it rarely exceeds 60 cm on the South Rim.

Havasupai is tucked away in a narrow, hard-to-find side canyon, while Navajos and Hualapai still dispute ownership of the lands along the rims of the abyss. In 1857 an army lieutenant travelling with a cartographic expedition to the canyon expressed his personal opinion of the place, with the pragmatism typical of military men: "The region is clearly of no value. We are undoubtedly the first and last group of white men to visit these places, for which no profitable use can be conceived". Over-hasty words... Each year nearly five million tourists crowd the rims, in the summer months especially, jostling to get the best view, the finest shots, the most spectacular light effects, or spend quiet hours looking out over what President Theodore Roosevelt described as "a superb spectacle that every American should see". But less than one visitor in a hundred dares test their strength against the giant, descending to the canyon floor. These most intrepid visitors are often a motley crowd: Japanese tourists — disciplined, eager and armed with telephoto lens —, city girls wearing elegant, high-heeled shoes, mad-keen hikers curved under the weight of huge, ergonomically designed backpacks which squeeze much that is useful (and plenty that is not) into a mere cubic metre; in common they have the same determination to make a journey backward through time as well as forward through space. The first part of the adventure is an easy downhill hike, but the going gradually gets harder (the temperature on the canyon floor can reach 43°C and signs at the start of the trails offer fair warning: "A hike in the Grand Canyon is a desert hike"); the return journey up the almost vertical incline can be hellish. But the rewards for so much effort make it all worthwhile: old abandoned mines, Indian ruins, rock engravings, the chance to pass through no fewer than six of the Earth's seven distinct climatic zones — from desert to artic/alpine —, encounters with numerous kinds of plant and animal life (fellow hikers included)

and — who knows — perhaps even an opportunity for introspection and self-discovery. There are two main trails for the adventurous descent into the canyon: Bright Angel Trail, first used by tourists in the late 19th century, and the steeper South Kaibab Trail. Both start from the South Rim, the most visited side of the park which offers twelve breathtaking vistas, including Grandview Point. Both lead to the legendary Phantom Ranch, a handful of cabins on the canyon floor which provide overnight accommodation (reservations needed months in advance); all its supplies are brought down by mule. The North and South Rims offer different kinds of canyon experience. The South Rim, approached from Grand Canyon Village, has motels, restaurants and museums on the geology of the canyon and the history of its indigenous populations; erosion has left it 366 metres lower than the northern edge, and its climate is drier. The North Rim is cooler and more moist, its vegetation more verdant; its scenery is characterized by the wild Walhalla Plateau and the North Kaibab Trail, its flora and fauna are more varied. Animal life and vegetation are abundant in the park. There are no fewer than 70 species of mammals, 250 of birds, 25 of reptiles and 5 amphibians. Here mule deer, desert cats, tamias, squirrels, wapiti deer, steller's jays and hermit thrushes live undisturbed, together with peregrine falcons and white-headed eagles. At higher altitudes white Colorado pines and trembling poplars flourish, lower down there are ponderosa pines, gambel oak and utah juniper and, in the depths of the canyon, spiny acacia and mesquite. Theodore Roosevelt made a major contribution to conserving wildlife in this unique, awe-inspiring habitat: in 1908 he established the Grand Canyon National Monument, eventually leading to the creation of the national park (in 1919). And the words of this United States president are surely the finest compliment paid to this great wonder of nature: "Leave it just as it is. It can't be improved on".

26-27 The amazing landscape of the Grand Canyon was created by the Colorado river, which has continued to wear away the rock walls of the plateau at the rate of 2 centimetres each century. The first white man to challenge the river's whitewater rapids was a trapper, James Ohio Pattie (1826), followed by Joseph Christmas Ives (in 1856) and subsequently by a one-armed Civil War veteran, John Wesley Powell who, with four boats and about ten companions, made a first successful journey down the Rio Colorado in 1869, and repeated the venture in 1872. The park is the habitat of about 70 species of mammals, including mule deer, desert cat and American red squirrel; there are also some 250 species of birds, among them the hermit thrush, steller's jay, wild turkey, blue grouse and virginia owl. One of the reptiles found in the canyon is the chuckwalla, an iguanid lizard which grows to a length of 40 centimetres.

CANYON COUNTRY

28-29 The satellite picture of the Grand Canyon shows an almost complete stretch of the winding course of the Colorado river which began to create this gigantic trench in what are — for geologists — fairly recent times: only 5-6 million years ago.

30-31 The first photos of the Grand Canyon were taken in the late 1800s by Jack Hillers who, in 1871, took over from E. O. Beaman as official photographer on John Wesley Powell's expedition to the Western territories. Also dating back to the end of the last century are the pictures taken by Timothy O'Sullivan (1871) and Frank J. Haynes (1880).

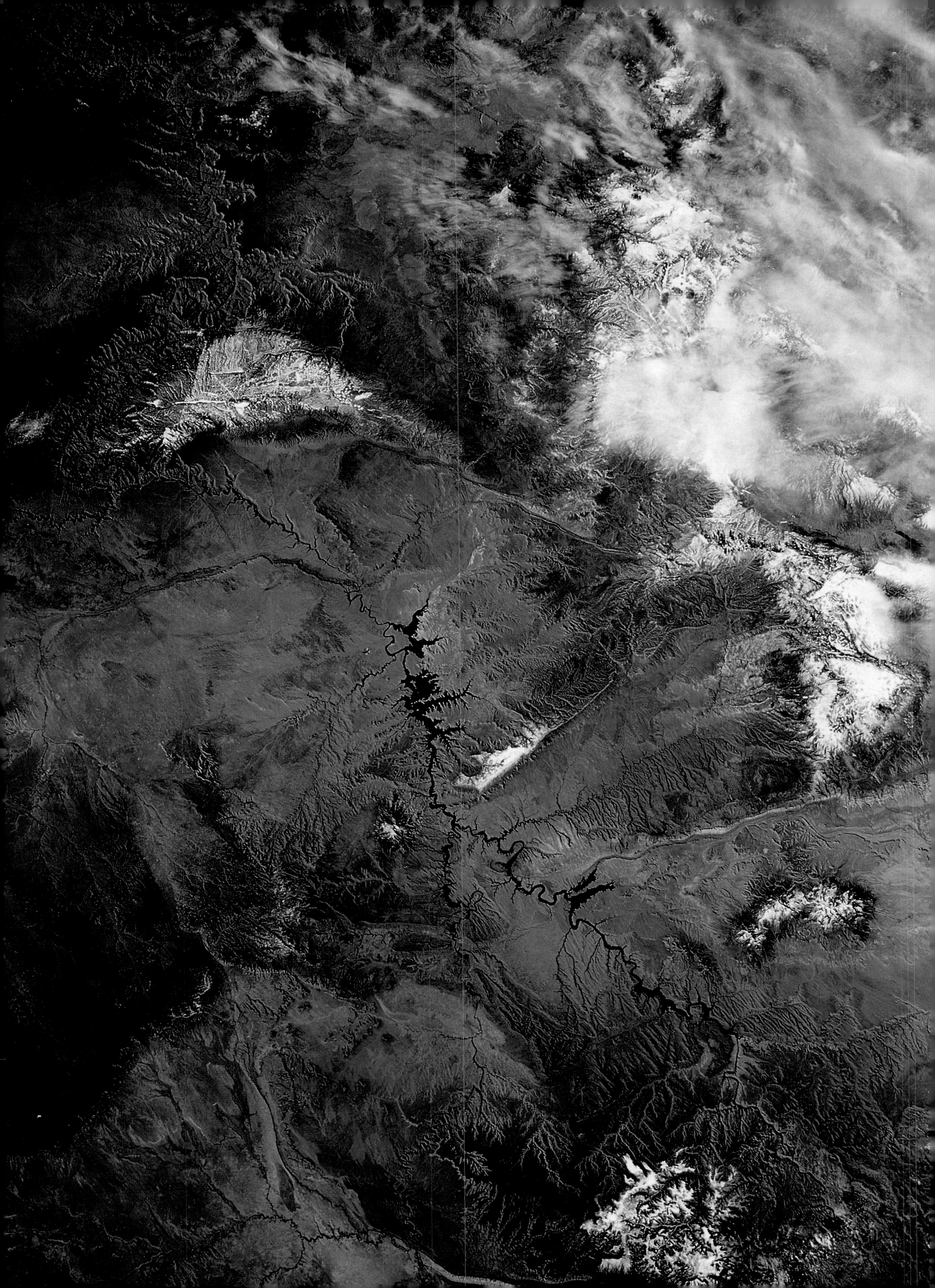

CANYON COUNTRY

32-33 The Grand Canyon National Park is 445 kilometres long, 30 kilometres wide at its broadest point and covers an area of almost 5,000 square kilometres; in certain places the canyon reaches a depth of almost 1600 metres. The first white men to come across the abyss were probably Spanish soldiers under the command of Garcia Lopez de Cardenas in 1540. They spent several days searching for a path along which to descend to the floor of the canyon, bed of the mighty Colorado, the second longest river in America: 3,219 kilometres.

34-35 Admiring the splendid landscape is, for some tourists, only part of the Grand Canyon experience. There are many more active ways of discovering the countless pleasures of the park. Among the most popular are trips aboard rubber rafts (along quiet stretches of the river or down rapids and waterfalls) or special high-bow crafts, for anything from a day to two or three weeks; another option is to ride on mule-back along the trails that descend to the canyon floor. For people who choose to go rafting down the white waters of the Colorado, Lava Falls present one of the biggest challenges. And for the adventurous (and physically fit), trekking along old Indian trails or paths made more recently by mining engineers — like Bright Angel Trail, first used at the end of the last century, or South Kaibab Trail (created by the National Park Service) — is a once-in-a-lifetime experience.

36-37 The Paiute Indians knew it as "the water that sinks into the ground"; for the Hopi it was "the descent into the bowels of the earth". Since time immemorial men have been drawn to the Grand Canyon and captivated by its splendour: paleo-Indians, Anasazi (first to farm in this area and build settlements, now ancient ruins), Hualapai, Havasupai (still living in a small side canyon) and Navajo. For all the Indian tribes who populated this region the Grand Canyon was a holy place. Deep in the gorge, about seven kilometres from the confluence of the Little Colorado and Colorado rivers, is Sipapu, believed by the Hopi to be the entrance to the bowels of the earth from which all living creatures come and to which they return after death.

38-39 President Roosevelt did much to ensure that the canyon, river and surrounding territory — the "inalienable heritage of all mankind" — remain unspoiled for future generations, first prohibiting hunting in 1906 and later establishing a national park in the area.

40-41 Deeply concerned about the state of health of the Grand Canyon, the park ranger service recently launched an SOS: the combined effects of pollution and hordes of tourists are putting the delicate natural equilibrium of the park at risk. The number of visitors per day reaches a peak of almost 27,000 in August, while the daily figure for the rest of the year is around 16,000. In the last ten years the number has doubled, from two million in 1984 to over four and a half million in 1994. And forecasts point to a figure of seven million by 2010.

MONUMENT VALLEY

This valley of awesome dimensions, its towering red stone cathedrals and gigantic buttes seemingly ablaze under the clear blue sky, has many tales to tell: Kit Carson giving chase to the Navajo Indians led by Manuelito, John Wayne galloping into the sunset — again and again —, the goldrush triggered by a phrase ("Go West, young man"), written in the New York Tribune in 1859, the memorable battle charge in 'Stagecoach', perhaps John Ford's finest film. This vast, barren plateau of Arizona, on the border with Utah, has provided a unique set for decades of Westerns and commercials: a backdrop so extraordinary as to seem unreal. It has also seen great tragedies. No half-measures can exist here: the valley may be overshadowed by threatening lead-grey thunder-clouds or inundated with light under a canopy of uniform blue. Which is why film directors like John Ford and Sergio Leone chose it as the setting — but also the protagonist — of countless stories of the legendary West which owed more to vivid imagination than to reality. And as their cameras rolled, they created a legend — the America of wide-open spaces and 'Injuns' — ready to be packaged and sold on the nostalgia market. No special effects are needed here: crystal-clear skies, gigantic boulders, dazzling colours, shimmering desert air, multi-coloured blankets woven by native Americans, silver jewellery... The stereotyped images created by cinema and now embedded in our minds fall far short of the reality of Monument Valley, where dimensions and space defy our wildest fantasies. "Watch out for animals for next 154 miles" warns a sign along one of the dirt roads that wind around the park and its monuments: what could better convey the unique feeling of boundless space that strikes even the most experienced globetrotter... And as we venture across this real-life film set, the actors of the old Westerns are still with us. For this is the traditional homeland of the Navajos, a small part of the reserve where descendants of the legendary Indian braves still live. There are just a few

42-43 When introducing himself he was brief and to the point: "John Ford's my name. I make Westerns". And Monument Valley made its real movie debut thanks to this great American director who used this 'set' for no fewer than seven films, most notably 'Stagecoach' in 1939. The fashion he started and others followed — among them Sergio Leone, Italian director of 'spaghetti Westerns' — continued well into the fifties and sixties with numerous other films, not all of them big screen greats. Here King Vidor filmed 'Duel in the Sun', here a young cowboy called John Wayne first appeared in an acting role he was to repeat for the rest of his life; here, in a scene reminiscent of Ford's own style, the recent adventure of Thelma & Louise came to its end. In Monument Valley a sign still indicates John Ford's Point, where the celebrated director took his favourite shots.

families of herdsmen here now, their modest homes almost always built at the side of 'hogans': traditional, circular, single-room dwellings, now used as accommodation for tourists. Freezing cold in winter and unbearably hot in summer, this arid, hostile land succeeded in halting the Spaniards and Mexicans but has had to submit to conquering tourists: in Monument Valley, with just a few dollars, they can buy a ticket to a dream, albeit short-lived. As with every film worth watching, there is a charge for taking the 27 kilometres. Valley Drive, a pitholed dirt road which crosses a bleak wasteland populated by towering buttes and mesas, the gigantic blocks of rock with steep sides and level surfaces that are the characterizing feature of park. The scenic side of the legendary West is presented, in an abridged version, in a tour with 11 stops, which takes in all that 'serious' tourists want and expect to see: solitary red sandstone monoliths, rock-ribbed sand dunes, sculptured gorges. This timeless, epic landscape, has come to symbolize America of yesterday and today, of reality and idealism, of history and fiction. Which is surely why the adventures of good-guy Tex Willer are situated here and why George Herriman chose it as the backdrop for his surreal cartoon tales about Krazy Kat. This park too owes its awe-inspiring scenery to the constant toil of wind and water; for millions of years they continued to wear down the surfaces of the enormous rents and fissures opened up in the Colorado Plateau. The giants of Monument Valley thus came into being and they are now such familiar sights as to be identified by name: Three Sisters, Totem Pole, The Thumb and many others. In the film 'The Searchers', after her son has been killed by the Comanches, Mrs Jorgensen finds the courage to say: "Some day this country's going to be a fine, good place to be. Maybe it needs our bones in the ground before that time can come". That time has now come but men still need to dream. In Monument Valley a sign marks John Ford Point, the exact spot where the great director most liked to set up his cameras. One step forward and you move from the world of reality to the realm of fiction.

44-45 Monument Valley is part of what is known as Navajoland, an immense territory famous for some of the most stunning scenery in the American West.

46-47 With Monument Valley within their reservation, the Navajo were to first to benefit from Ford's films. While shooting 'Stagecoach', for instance, Ford spent about $200,000 in the valley. As a token of thanks, the Navajo named him Natani Nez, Great Soldier, and made him an honorary chief of their tribe.

48-49 The vast landscape of Monument Valley, dominated by towering red mesas like Castle Rock and Big Chief, was bound to appeal to a man like John Ford. "I like fresh air, wide open spaces, mountains and deserts — he used to say — Sex and the like... are not for me".

50-51 Nowadays Westerns are in decline but admen making TV commercials and adverts have been "crowding" the valley, which obtained national park status in 1958. And tourists gladly pay the entrance fee to make the dirt track journey (a roundtrip of 27 kilometres) to eleven breathtaking vista points.

52-53 Summer is the season preferred by tourists to Monument Valley even if here, as in all semi-desert places, daytime temperatures can soar to 35-40°C.

54-55 Scattered across the valley are incredibly tall buttes and colossal mesas: characteristic, isolated elevations with sheer sides and level tops created by the erosive action of water, ice and wind. The valley is home to a small Navajo community: as few as twenty people, mostly herdsmen. In the sixties, thanks to coal mines and oil extraction, the Navajo were the richest tribe in the USA. Income from these activities was invested to improve living conditions in the huge reservation of which Monument Valley is part.

56-57 The rainy season in the valley is generally concentrated in August and September. In the winter the first snows fall on the mesas as early as November, and the temperature can drop to -10°C.

58-59 As in many other national parks of America, the bizarre rock formations of Monument Valley have acquired fanciful names inspired by their shape, for instance Castle Rock, Totem Pole and Three Sisters, considered outstanding highlights of the park.

60-61 The awe-inspiring structures scattered across the barren wastes of Monument Valley are the outcome of a process of erosion that has lasted for more than 25 million years. In the course of millenia water and wind eroded the surfaces of huge rents and fissures in the plateau that once covered this part of southwestern America, caused by the upward thrust of the Earth's crust due to pressure from its interior.

62-63 Cowboys and Indians, as well as amazing rock formations, are fundamental ingredients of the 'Wild West' scenery served up by what is generally considered the main tourist attraction of Navajoland.

64-65 As for the Grand Canyon, the magnificent landscape of Monument Valley can be seen at its finest from above: near Oljato Trading Post or Monument Valley Hospital, it is easy to find pilots of small planes or helicopters who — at a price between $ 50 and 100 — will let tourists enjoy an unforgettable bird's eye view of the scenic wonders on the ground.

LAKE POWELL

The Colorado river has unquestionably played the leading role in the geological history of America's West since without this one-time wild and unruly monster the spectacular Grand Canyon would never have existed. Further credit goes to the Colorado for having helped create one of the most splendid sites in the whole USA: Lake Powell, in the states of Utah and Arizona. Reaching a length of over 300 kilometres and with a tortuous shoreline of over 3,200 kilometres, the lake came into being after completion in 1963 of the huge dam — built upstream from the Grand Canyon — which soon tamed the previously fast-flowing Rio Colorado. Standing 215 metres high, the dam was constructed at the narrowest point of the huge Glen Canyon which, now submerged by the waters of the reservoir, has been turned into one of the most attractive lakes in the South-West. While the dam has undeniably provided local populations with water and electricity, vociferous environmentalists claim it has spoiled the natural beauty of the canyon. It may be true that the dam has not enhanced Glen Canyon but it has certainly made a big contribution to the fortunes of Page, a small town on the shores of Lake Powell. It has rapidly become a highly popular resort for water sports: swimming, water skiing, rafting, sailing, windsurfing and fishing. Vacationers can take trips on motorboats, rent houseboats and explore at leisure hundreds of winding ravines, tiny canyons and deserted beaches as well as historic remains of Indian settlements. Few places on Earth can have been the subject of more photos than Corkscrew Canyon, just a few miles from Page; after ascending to Romana Mesa in the north of the canyon, at an altitude of 2000 metres, the many wonders of the American West can be summed up in one single, glorious, panoramic view. Another 'must' is an excursion to the nearby Rainbow Bridge National Monument, venerated by Navajo Indians as a petrified rainbow: standing 90 metres high, it is one of the world's largest natural arches.

66-67 With a length of almost 300 kilometres and some 3,200 kilometres of shoreline, Lake Powell is one of the largest artificial lakes in the USA and definitely the most spectacular. On its banks stands the town of Page: its urban boundary coincides with the edge of the huge Navajo reservation, which occupies the whole northeastern quarter of Arizona and also enters Utah, Colorado and New Mexico.

68-69 and 70-71 Built from 3,825,000 cubic metres of concrete, Glen Canyon dam stands 215 metres high and 474 metres wide, linking the sheer cliffs of the gorge. The combined work of man and nature, it offers an awesome sight. The lake and its environs, covering hundreds of thousands of hectares, have been designated Glen Canyon National Recreation Area. Teddy Roosevelt was strongly in favour of the scheme to construct dams in this area, as a result of which the Colorado river became — as writer Philip Fradkin put it — "the most abused, contended, politically used and newsworthy river in the country, if not in the world".

72-73 There was a long and bitter controversy over the dam built in Glen Canyon, to which the reservoir owes its existence. Commenced in 1956 and completed in 1963, its construction was vehemently opposed by ecologists, who fought hard to save from a watery grave a canyon no less awesome than the most eminent examples to be found on the Colorado Plateau, including the Grand Canyon itself; they also feared the effects of limiting the flow of the once-mighty Colorado river. In spite of its controversial history, Lake Powell is an extraordinary sight, an amalgam of the many scenic features the American West is famous for: canyons, deserts, glowing red and pink sandstone, sand dunes and crystal-clear water.

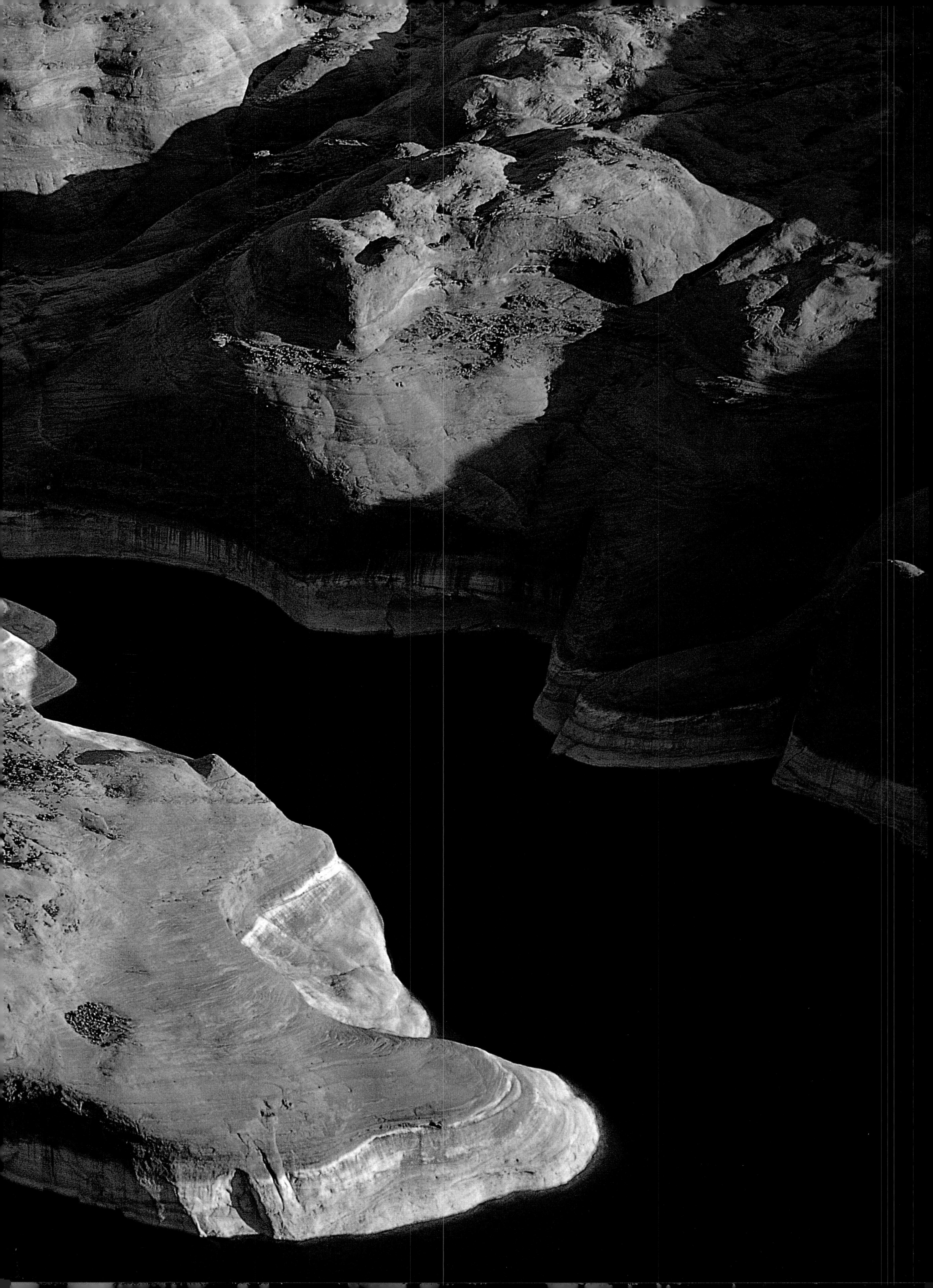

82-83 The panorama in Bryce Canyon National Park, in Utah, is characterized by sculpture-like siltrock pillars, in shapes recalling castles, towers, skyscrapers and Indian temples.

84-85 In summer the temperature in the park can easily reach 40°C, dropping to 10°C at night, but the average temperature in this season is a comfortable 20°C, making it an ideal time to visit and explore the park. In spring and fall however Bryce Canyon is more sparsely populated with tourists.

86-87 The fascination of the rocky depressions of Bryce Canyon lies not only in the bizarre shapes created by natural weathering: it also stems from their subtle colours which range from blue to cream, pink to orange, white to red and even purple, and take on a magic glow in the light of the setting sun. The park's most popular attraction is Bryce Amphitheater, which can be admired from several viewpoints.

were deposited on the bottom of the lake and eventually turned to rock: the limestone and sandstone of the Wasatch mountain range. Sixteen million years ago upward movements of the earth's crust forced this range to the surface; the lake disappeared and the crust broke up into huge blocks, plateaus like Paunsaugunt, on the east side of which the erosive power of water created the monumental landscape now known as Bryce Canyon. This weathering process still continues today at an incredible speed, in geological terms: every 65 years the canyon rim recedes 3 centimetres. Gulliver's Castle, Thor's Hammer, The Cathedral, Silent City, Hat Shop, Wall of Windows: just as nature revealed her infinite creative range, so men have stretched their imagination to its limit to find names for them. Of the many trails offered by the park, Rim Trail, which winds around the edge of the Bryce Amphitheater, is the easiest and most scenic; Fairyland Loop Trail, starting and finishing at Sunrise Point or Fairyland Point, offers spectacular views of some of the most outstanding rock formations — Chinese Wall, Tower Bridge, Ruins of Athens — and passes twisted old bristlecone pine trees (one reckoned to be 1800 years old). Perhaps the most splendid lookout is Bryce Point, where there is a spectacular view over the pillars of the amphitheatre. More magnificent scenery can be seen from Rainbow Point and Yovimpa Point, stretching away towards the Grand Canyon, the last of the rocky "terraces" which, from Bryce, drop down across Zion National Park to the Kaibab plateau, joining Utah and Arizona. With an ambient noise level of 14 decibels, Bryce Canyon is one of the most silent places in the whole of America. A small contribution to increasing this threshold is made by the park's fauna: prairie dogs, tamias or chipmunks, mule deer, badgers, short-horned lizards, pumas and wapitis along with no fewer than 164 species of birds. At lower altitudes the amphitheatres are surrounded by scrub pines, Utah juniper, artemisia and Mexican Cowania, replaced above 2700 metres, in the southern part of the park, by conifers like the white Colorado pine. Ebenezer Bryce's cow could have done far worse than get lost in such a place!

88-89 The heaviest snowfalls of the winter are generally between December and March. According to the findings of recent research, the freeze-thaw cycle takes place here no fewer than two hundred times a year. And yet because of the permeable surface of the rock, there is never a lot of water to be found in the park. The trees of the plateau have adapted well to this potential problem: when snow and ice melt, they store moisture in their roots as a reserve for the hot, dry summer ahead. In spite of the snow, winter visitors to Bryce Canyon can tour the rim of the canyon wearing sealskins and snow-shoes or cross-country skis, supplied by the Visitors Center. For safety reasons, however, only experienced climbers (with permits issued by the park's ranger service) may descend onto the floor of the rocky amphitheatre.

SCENIC WONDERS OF UTAH AND ARIZONA

Canyonlands, Natural Bridges, Petrified Forest: here too rock has been shaped into huge arches, eroded to form deep gorges, twisted into pillars and pinnacles, crushed to fine sand, cracked like the skin of a lizard or as towering red mesas. This barren wilderness has been crossed by adventurers, blue-jacketed soldiers and black-apparelled Mormons. It is the mysterious and mystic land of Anasazi and Navajo tribes, ghost towns, cliff dwellings, inaccessible canyons, glittering silver jewellery and polychrome terracotta wares. It is also the timeless world of the Petrified Forest National Park in Arizona, with the silicified remains of pine trees that grew here 225 million years ago. This arid tableland was once a fertile, tree-rich plain. After its trees fell and were covered by mud, lack of air turned their organic matter into coloured crystals of first silica, then quartz. As the result of geological upheaval and constant erosion by wind and rain, the tree trunks, stumps and tiny fragments of wood eventually returned to the surface, as a petrified forest. Ruins of ancient 'pueblos', or cliff dwellings, and paintings on rock walls are all that remains of populations who once inhabited these parts and had abandoned them before 1540 when the Spanish arrived here. Discovered in the mid-1800s by an American army officer, the area designated in 1962 as the National Park includes 1000 hectares of the multi-coloured badlands of the Painted Desert; the park has a small museum with exhibits explaining the process of wood petrification. Canyonlands National Park, the largest in the state of Utah, has a seemingly extraterrestrial landscape, a rocky desert set with ornamental pillars, tablelands and deep canyons. Across this vast wilderness (1,365 square kilometres) flow the Colorado River and its tributary, Green River, which start life in the mountains of Wyoming and Colorado. The rivers converge at the heart of the park, dividing it into three separate parts: in the north is Island in the Sky, a triangular tableland furrowed by numerous canyons, some plunging to a depth

90-91 The multi-coloured hills of the Painted Desert and vividly coloured petrified tree trunks are the main attractions of the Petrified Forest National Park. Almost 225 million years old, the remains of these primordial trees — silicified after being buried beneath layers of sediment and sand — eventually returned to daylight after movement of the Earth's crust changed the face of the plateau. Among the park's amenities is the Rainbow Forest Museum where the process of wood petrification is explained in detail.

92-93 The northern section of the park includes part of the Painted Desert, a stunning natural scenario comprised of hills and tablelands formed from sandstone, schist and clay, with colours ranging from flaming reds to dusty brown and brilliant orange.

94-95 Preserved in the Canyon De Chelly National Monument are the ruins of about sixty settlements abandoned by peoples who lived in this area long ago, between 350 and 1300 AD, from the Anasazi to the Hopi, Pueblo and Navajo. Among the main attractions of this important archeological site, formed of three deep gorges created by the Chinle River, are two sites 'wedged' into deep crevices in the rock: the White House (left and above), the only one explorable without a guide (compulsory for all other visits in the canyon) and Mummy Cave Ruin, one of the most awe-inspiring constructions in the park (below).

of 800 metres; to the west is The Maze, the most remote section; to the east is The Needles, with its collection of pillars and pinnacles — like the assortment of bizarre eroded shapes along Joint Trail — and famous arches: Angel Arch, The Castle and Fortress Arch. Highlights of the park include stunning views from Grand View Point and Maze Overlook, the spectacular Cataract Canyon, a rafting paradise, and the ruins and rock inscriptions left, in the Needles District, by Fremont and Anasazi. Until the early 1800s when these lands were used as pastures by white settlers (and as a hiding-place by Butch Cassidy and the Sundance Kid), these early native inhabitants were the sole occupants of the region besides chipmunks, coyote, mule deer, snakes and lizards. Only two blacktopped roads lead into park, from the Island in the Sky and The Needles entrance points. Also in Utah, and relatively close to Canyonlands, is the Natural Bridges National Monument: three huge natural bridges (the largest 67 metres high with a span of 82 metres) which, as the rock formations get thinner, show different stages in the erosion of their Cedar white sandstone. Several steep trails lead to the base of the bridges, which have Hopi names (Sipapu, Kachina and Owachomo) and were first discovered by whites only in 1883. The many ruins scattered in this isolated, uninviting territory are yet another testimony to the Anasazi civilization. Like the mysterious Anasazi of Mesa Verde or the ill-fated Navajo of the Canyon De Chelly, these peoples surrendered to their destiny, even before the native Indians' struggle against the whites was lost. The Canyon De Chelly National Monument in Arizona, established in 1931, is situated in the largest Indian reservation in the USA, home to over 200,000 Navajo, the most numerous native Indian tribe in North-America today. It was also the scene of the most tragic Navajo defeat. Here, in January 1864, after a ten-year-long campaign to subdue the Navajo, sublieutenant Kit Carson and his 300 men laid seige to their old enemy and starved them into defeat. After burning the Indians' crops, destroying their hogans and well-tended peach plantations in the Canyon De Chelly, slaughtering their herds and

96-97 The name De Chelly comes from a Spanish corruption of 'tsegi', a Navajo word meaning "rock canyons". The largest of the three gorges of this National Monument is the Canyon del Muerto, so-called since 1882 when an expedition organized by the Smithsonian Institution discovered prehistoric burial-places here.

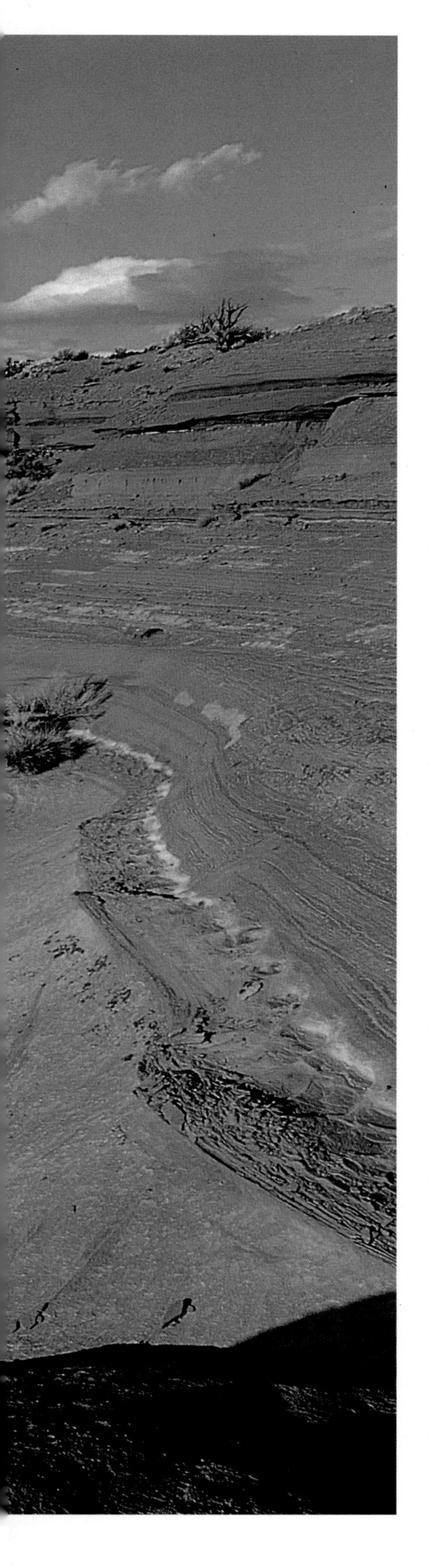

horses, Carson won the day. The 8,000 survivors set out on the Long March — almost 500 kilometres — to Bosque Redondo in New Mexico, where a quarter of them eventually died of hunger and disease; the remainder were allowed to return only in 1868. The canyon had never brought fortune to the Navajo: earlier, in 1805, they had suffered defeat at the hands of Antonio Narbona, the Spaniard later appointed governor of the province of New Mexico. Located some 400 kilometres east of Grand Canyon Village, the Canyon is formed of three deep gorges carved by the waters of the Chinle River. Clinging to rocky cliffs or sheltering in deep crevices are some sixty ancient ruins (dated 350-1300 AD) of settlements built here by native Indians: Anasazi, Hopi, Pueblo and Navajo. Two roads cross the park and access is controlled by the National Park Service, with guided tours only. The sites of greatest interest are the White House ruins — the only place that can be toured freely —, Antelope House, Standing Cow (with Navajo graffiti), Mummy Cave and Spider Rock, a freestanding sandstone pinnacle over 240 metres high. Discovered only in 1888 by two roaming cowboys, Mesa Verde, in Colorado, offers more extraordinary evidence of the ancient Anasazi peoples whose civilization originated almost 2,000 years ago, only to vanish — perhaps when they moved to southern Arizona and New Mexico — around the 13th century. The forty or so archeological ruins contained in this National Park are among the best preserved in America. Dating back to 500 A.D., they are multi-storied cliff dwellings — or 'pueblos' — built from sandstone blocks held together with clay in the crevices of rocky cliffs, hundreds of metres from the valley floor. In the vicinity of the ruins is a museum that, with dioramas and displays of artifacts, reconstructs the life of this ancient civilization, at its height around the year 1100. Among the most scenic ruins of Mesa Verde are Cliff Palace, containing no fewer than 200 rooms, almost perfectly camouflaged against the rock wall, and Spruce Tree House, a settlement of 114 dwellings beneath a huge, overhanging cliff.

98-99 Created by the river of the same name, Paria Canyon in Arizona is at the centre of the Paria Canyon-Vermilion Cliff Wilderness Area, established in 1984. Government-owned and operated by the Bureau of Land Management, this area of 44,000 hectares is in many respects still completely unspoiled.

CANYON COUNTRY

100-101 From the top of Paria Canyon visitors are greeted with breathtaking vistas of rounded mesas and far-distant horizons; in the course of time water and wind have hollowed out deep depressions in the canyon walls, exposing the mineral veining typical of the very oldest rock formations.

102-103 Canyonlands National Park covers an area of 1,365 square kilometres, surrounding the confluence of the Colorado river and its tributary, Green River, which divide the park into three separate sections. With a dozen or more layers of different rock, the walls of the canyons created by these rivers offer an amazing overview of some 300 million years of geological history.

A secure hiding place in the 1800s for bandits Butch Cassidy and the Sundance Kid, Canyonlands is one of the wildest and most inaccessible parks of the USA. Used as pastureland by white settlers only since the early 19th century and explored in 1869 by Major John Wesley Powell, the first man to successfully brave the mighty Colorado and Green rivers, the area became a national park only in 1964.

104-105 Although access to some parts of Canyonlands National Park presents problems, it can be toured on foot, by car or off-road vehicle and even in a boat or canoe, along the Green and Colorado rivers. Nevertheless each of its three areas — Island in the Sky in the north, The Maze in the west, The Needles in the east — has trails which allow visitors to make unforgettable excursions in this almost entirely virgin wilderness.

106-107 Island in the Sky is best known for its yawning canyons, some plunging to a depth of 800 metres; in The Needles section there are spires, pinnacles, arches and other weird rock formations shaped like mushrooms, towers and giants. The Maze is the most remote and wild area of the park, scattered with canyons, mesas and intriguing eroded rock sculptures.
The most impressive views are afforded by Green River Overlook (Island in the Sky), Pothole Point (Needles) and Maze Overlook (Maze).
The ancient inhabitants of the area have left traces of their presence: ruins scattered in various parts of the park, and the pictographs of Newspaper Rock.

108-109 Mesa Verde in Colorado has some of the best preserved rock and cave dwellings in America. Discovered by two cowboys in 1888, it contains dozens of pueblos and Indian settlements that have survived since the 6th century A.D. and thousands of prehistoric sites dated between 550 and 1270 A.D.

100-111 With its three immense natural arches, discovered by whites only in 1883, the Natural Bridges National Monument in Utah provides evidence of the various stages in the erosion by wind and water of Cedar-formation sandstone.

CANYON COUNTRY

112-113 Close to the San Juan River, on the Utah-Arizona border, is the Valley of the Gods, a very popular destination among tourists in search of intriguing rock formations. Its popularity stems in part from Highway 261 which overlooks the valley and is an ideal spot for taking wonderful panoramic photos. Occupying an area of area of 350 square miles, the valley is formed of a deep depression, situated on the edge of the immense Navajo reservation.

114-115 Huge buttes and mesas, pinnacles and towers of rock created by the eroding action of the water and wind: geographical features which make the Valley of the Gods a look-alike of Monument Valley (from which it is separated by the San Juan River). Thanks to John Ford and his films, the Western myth lives on in this world-famous landscape.

ARCHES NATIONAL PARK

After the Grand Canyon's dizzying cliffs and Bryce Canyon's spires — like spiky dreadlocks — we come to the bizarre formations of Arches National Park, established in 1971: an amazing collection of arch-shaped "sculptures", seemingly abandoned by croquet-playing giants. These incredible natural wonders populate an area of about 300 square kilometres in the state of Utah, 8 kilometres north of Moab: free-standing arches, double arches, pillars topped by massive boulders, an astounding landscape of amazing rock shapes to walk under, climb up, ride over or hang from. This face of America is perhaps not instantly recognized like Monument Valley. Its scenery may not be as dramatic, awe-inspiring and sensational. But it is still the America of superlatives and hyperbole, with a fascination few can resist. Drop after drop, gust after gust, water and wind have sliced and then hollowed their way through millions of cubic metres of rock, creating a collection of amazing natural bridges, freestanding arches of still changing shape. Some are so thin that they look dangerously fragile, others so huge and overwhelming that even the first explorers of the area – men hardened by gruelling experiences – described them as a moving sight. A few have become symbols: pictures of Delicate Arch, for instance, have reached every corner of the globe as the spectacular backdrop for adverts or illustrations on countless glossy travel magazines. Arches' fama worldwide has made it one of the most visited parks in America and to be photographed astride a slender natural bridge is something millions of tourists dream of. In this majestic but barren desert setting, at a height of between 1200 and 1700 metres above sea level, Arches has numerous easy excursions to delight visitors. The park is crossed by asphalt roads (covering almost 30 kilometres), dozens of dirt roads, and many designated trails. The arches,

116 Arches National Park was established in 1971 to the north of Canyonlands National Park. The Colorado river marks one of its boundaries.

117 The beautiful Delicate Arch is undoubtedly the most celebrated of the approximately 1,500 listed natural arches in Arches National Park. Standing 13 metres high, with a span of 10 metres, this well-known rock formation overlooks the imposing landscape of the La Sal mountains.

spanning from 90 centimetres to nearly 100 metres, provide an excellent gymnasium for climbers and some of them have seen daring exploits, recorded in the annals of U.S. mountaineering and free climbing. The process of erosion which eventually created these elegant arches began over 300 million years ago when a primordial inland sea evaporated leaving behind an uneven salt bed, in places 3000 metres thick, on this part of the Colorado Plateau. The salt crust was eventually buried under eroded sediment from the surrounding mountains and, in time, the detritus compacted into heavy rock, several hundred metres thick. The uplift of the plateau and the twisting and crumbling of the salt under the weight of the rock created a series of fissures and rents and brought to the surface red and yellow Entrada sandstone which had been lying deep beneath the Earth's crust for 140-150 million years. Further movement of the rock strata far below the surface eventually caused parallel rents. Water,wind and ice were thus able to "chisel" their way through the sedimentary rock, creating first a series of layered structures, then recesses and holes which grew in size until they formed the arches which are now the characterizing feature of this landscape. And these same forces of nature are still at work today. The eloquent names given to the Arches' rock formations show that there is no limit to the ideas these shapes inspire: Three Gossips, for instance, or Baby Arch, The Organ, Tower of Babel, Elephant Parade. The undisputed stars of the show are Ribbon Arch — no more than 30 centimetres wide at its narrowest point — and the world-renowned Delicate Arch (13 metres high, 10 metres wide), a spectacular sight against the imposing backdrop of the La Sal mountains, especially at sunset. Of the park's trails, the challenging Devil's Garden Trail boasts no fewer than 64 arches, including Landscape Arch (89 metres), the longest natural rock arch in the world. With four of the park's most famous arches to its credit — Double Arch, Turret Arch, North and South Windows — the Windows

118-119 Situated on one of the tablelands of the Colorado Plateau, the park boasts an impressive collection of layered rocks, sheer-cliff canyons and arches that have been a training-ground for generations of American mountaineers and free-climbers. The most popular arch among climbers is the slender Landscape Arch; with its span of 88.70 metres, it is believed to be the longest in the world. It was first climbed in 1939 by Philip S. Miner, a 19-year-old member of the Wasatch Mountain Club of Salt Lake City. Another young climber was not so lucky: just nine months later he fell to his death when climbing a ridge on the north face. All intending climbers are advised to consult the park rangers about the solidness of any rocks they plan to tackle.
In many places this sandstone does not provide a secure hold for pitons and may be friable and slippery. Climbing on rocks which bear engravings and pictographs left by the Anasazi and Fremont Indians who used to hunt in this region between the 11th and 14th centuries is, of course, prohibited.

Section, at a distance of 15 kilometres from the Visitors Centre, is most definitely not to be missed (and the same goes for Balanced Rock, a towering slickrock pillar with a huge "head" which is a miracle of natural equilibrium). Not surprisingly, with temperatures reaching 40°C in summer, -8°C in winter and often changing brusquely as day turns to night, this rocky desert had little appeal to populations who came this way. Petroglyphs on stone walls in the park show that the Anasazi, Fremont and Ute wandered this area at different times through the centuries, but they came here to hunt rather than to establish permanent dwellings. The temptation to leave a trace behind was also too great for a certain Denis Julien, perhaps travelling along the Old Spanish Trail with a wagon train of white settlers: his name and the date — June 9, 1844 — are carved on one of the Arches' rocks. And yet even in this wilderness, as in Bryce Canyon, someone took it into his head to raise cattle. John Westey Wolfe, a Civil War veteran from Ohio, arrived here with his son Fred in 1888, possibly more in seach of solitude than of grazing land. And he must have found it in this godforsaken place, for he remained here until 1910. His dilapidated ranch remains to this day — now one of the Park's attractions and a tangible tribute to the determination and courage of the early settlers of the West. Tourists may have the run of the park all day long but, as dusk falls, the animal population comes into its own. This area is the natural habitat of nearly 200 species of birds, including ravens and red-necked vultures, white pumas, grey foxes, coyotes, mice, mule, deer, black-tailed jack rabbits, rat kangaroos and toads continue their daily struggle to survive amid the scrub pine and juniper which cover almost half the park. Single-leaf ash emerges from cracks in the rocks while bushes of thorny coleogyne thrive in this sandy soil. The stunted vegetation highlights the dryness of this desert wilderness where the scent of artemisia hangs in the air, contributing to the heady sense of real adventure that Arches National Park still conveys.

120-121 Although little rain falls in Arches National Park (about 22 centimetres per year), water rather than wind was the primary cause of the erosion that produced rock formations like the stunning Double Arch (top right). Rain first created fissures and gullies in the sedimentary parts of the rock; the lower layers of the solid rock left standing subsequently weakened and part of the walls subsided, but the arches of solid rock stood firm and have survived to the present day.

122-123 In Arches National Park, which covers about 300 square kilometres, are other sandstone sculptures besides the arches after which it is named: pillars, pinnacles and vertical structures that recall organ-pipes. Every year no fewer than half a million tourists visit the park, which has a 30-kilometre asphalt road leading to its most impressive natural wonders.

124-125 The little vegetation in the park consists essentially of cowania mexicana, Utah juniper and scrub pine but the single-leaf ash also thrives here (its tiny coriaceous leaves, as well as its small size, enable it to store the moisture needed to survive in this desert climate). Where the soil is sandy, blackbrush (Coleogyne Ramosissima of the Rosaceae family), artemisia and green ephedra grow, while vast stretches of land are covered with prickly pear cactus.

126-127 Situated just a few kilometres north of Canyonlands, the rock formations of Arches National Park offer extraordinary examples of howthe passage of time and erosion by wind and water have changed the face of our planet.

128 The timeless, mythical landscape of Monument Valley — where the few Indians who survived the Canyon De Chelly massacre sought refuge, under the command of Manuelito — is a compilation of all our visual expectations of the legendary West.

CANYON COUNTRY

ILLUSTRATION CREDITS

All photographs by Antonio Attini except the following:
NASA: pages 28, 29